A BOOK OF

WEATHER CLUES

COMPILED AND
ARRANGED BY

DIANE KAISER

STARRHILL PRESS
Washington & Philadelphia

Published by Starrhill Press, P.O.Box 32342, Washington, DC 20007

We gratefully acknowledge the kind permission given us to reprint the
following: From "Of the Surface of Things" and "The Comedian as the
Letter 'C'" in *The Collected Poems of Wallace Stevens*. Copyright © 1968 by
Wallace Stevens. Reprinted by permission of Alfred A. Knopf, Inc. Excerpt
from "Sido" in *Earthly Paradise* by Colette. English translation by Herma
Briffault, copyright © 1966 by Farrar, Straus and Giroux, Inc. Reprinted by
permission of the publisher. Specified excerpt (text only) from *Charlotte's
Web* by E.B. White. Copyright © 1952 by E.B. White; text copyright renewed
1980 by E.B. White. Reprinted by permission of Harper & Row, Publishers,
Inc. "Memorize Your House" by Claire Siegel. Copyright © 1986 by Claire
Siegel. Reprinted by permission of the author. From *The Member of the
Wedding* by Carson McCullers. Copyright © 1946 by Carson McCullers;
copyright renewed © 1974 by Floria V. Lasky. Reprinted by permission of
Houghton Mifflin Company. From *The Witches of Eastwick* by John
Updike. Copyright © 1984. Reprinted by permission of Alfred A. Knopf, Inc.
From "Preludes" in *Collected Poems 1909-1962* by T.S. Eliot. Copyright
1936 by Harcourt Brace Jovanovich, Inc.; copyright © 1963, 1964 by T.S.
Eliot. Reprinted by permission of Harcourt Brace Jovanovich, Inc. From
Chicago Poems by Carl Sandburg. Copyright 1916 by Holt, Reinhart &
Winston, Inc.; copyright renewed 1944 by Carl Sandburg. Reprinted by
permission of Harcourt Brace Jovanovich, Inc. From "Two Tramps in
Mudtime" in *The Poetry of Robert Frost*; edited by Edward Connery
Lathem. Copyright 1936 by Robert Frost; copyright © 1964 by Lesley Frost
Ballantine; copyright © 1969 by Holt, Rinehart and Winston. Reprinted by
permission of Henry Holt and Company. The cloud photographs were
supplied by the National Oceanic and Atmospheric Administration and are
reproduced here with their very kind permission. The drawings for this
book were adapted by Judith Curcio from Jennie Day Haines' *Weather
Opinions*, published in 1907 by Paul Elder and Company.

Special thanks to Bob and Micki Ravitz. For friendship.

Library of Congress Cataloging in Publication Data

A Book of weather clues.

    1. Weather—Folklore. I. Kaiser, Diane, 1946-
QC998.B62 1986   551.6'31   86-90363
ISBN 0-913515-11-6

Published simultaneously in Canada by Fitzhenry & Whiteside, Toronto.

Printed in the United States of America

First edition
   3 5 7 6 4 2

*For my Father
who taught me
which way
the wind blows.*

Nothing that is can pause or stay;
The moon will wax, the moon will wane,
The mist and cloud will turn to rain,
The rain to mist and cloud again,
Tomorrow be today.

Henry Wadsworth Longfellow
from *Kéramos*

Table of Contents

One Little Window

How many times have you been caught in a summer storm without an umbrella or a raincoat? Or in a snowstorm without boots? Have you ever wished for an extra jacket or wondered why, on the hottest, muggiest day of the year, you're wearing as many layers as an Egyptian mummy?

You may not know it but there are weather clues all around you.

Look, of course, at the clouds and test the wind, but take a look too at your cat, coffee cup, dandelions and spider (if you're lucky enough to have one around the house). Is your hair frizzing up? Have you had a restless night? Are the mosquitoes particularly ferocious? How's your appetite? If you live in a city, watch the pigeons.

This *Book of Weather Clues* tells you what to look for, and where to look, to put together the puzzle pieces of the day's weather.

Here are thoughts and observations from sailors, poets, farmers, inventors and scientists. Some of these clues go way back to many years ago. See if they work for you.

Included are pages for you to record your own observations, month by month. Did it snow early this year? When did you see the first robin? Did the groundhog see his shadow in February? Were the daffodils blooming, full-force, in April? How long did the July heat wave last?

This book won't make you a meteorologist, but chances are you won't be surprised should the sky turn black and the thunder start to rumble.

All it takes is a good look out one little window and your *Book of Weather Clues.*

Hopefully, your predictions will be a bit more on the mark than Mr. Samuel Langhorne Clemens' forecast for New England given to the New England Society on December 22, 1876:

> "Probable nor'east to sou'west winds,
> varying to the southard and westard
> and eastard and points between; high
> and low barometer, sweeping round
> from place to place; probable areas of
> rain, snow, hail and drought,
> succeeded or preceded by earthquakes
> with thunder and lightning."

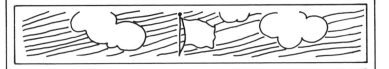

In my room, the world is beyond my understanding;
But when I walk I see that it consists of three or
    four hills and a cloud.

Wallace Stevens
from "Of the Surface of Things"

CLOUD CLUES – Fair Weather

The higher the clouds, the finer the weather.

Cumulus
Detached heaps of brilliant white clouds—the ones you made into all kinds of creatures as a child. Unless they grow tall and menacing, these clouds usually bring fair weather.

Cirrus
High, feathery filaments of ice crystals brushing the tops of the heavens. These delicate, detached tufts look like painters' brushes, or long commas, or mares' tails. Cirrus usually marks the start of a warm front.

Basics:
Cirrus(o-): high altitude clouds
Altus(o-): middle altitude clouds
Cumulus(o-): clouds in heaps
Stratus(o-): clouds in layers
Nimbus(o-): implies rain

CLOUD CLUES—Change

A dappled sky, like a painted woman, soon changes its face.

Cirrostratus
High clouds like thin skimmed milk spilled across the sky. They're thin enough for the sun to shine through and often precede a warm front. Look for rain in 24 hours.

Cirrocumulus
High, thin sheets of cloud looking like ripples on a lake bottom or the scales on the back of a fish. This is the famous "mackerel sky."

Mackerel sky / Rain by and by.

CLOUD CLUES—Rain

Dark clouds in the west at sunrise indicate rain on that day.

Altostratus
A bluish or grayish flannel sheet of clouds through which the sun can appear as through frosted glass. This sky usually means rain in 6 to 12 hours and often accompanies a cold front.

Altocumulus
Small rounded clouds of white and gray, grouped together in rolling sheets. These clouds look like the woolly backs of sheep or a full-blown head of cauliflower. Rain will come in 10 to 15 hours.

Red sky at night / Sailors' delight.
Red sky at morning / Sailors take warning.

CLOUD CLUES—Rain

Rain before seven / Clear before eleven.

Stratus
Low, gray, misty sheets
of cloud, sometimes low
enough to be fog. These
clouds almost always
mean precipitation.

Stratocumulus
Low, heavy, dark waves
of cloud with rounded
masses. These often
precede a cold front.
Watch out for
thunderstorms and
winds.

Ground fog in the early morning, following a clear
night, portends a fine day.

CLOUD CLUES—Rain

When clouds appear like rocks and towers,
The earth's refreshed by frequent showers.

Cumulonimbus

The great thundercloud, it looms high in the sky, grows an ominous anvil-shaped top and is as dark and threatening as shadows in a back alley. There will be thunder and lightning and heavy rain forthcoming.

Nimbostratus

A thick, dark and diffuse cloud layer. These often come from altostratus clouds growing denser and lower as they drop rain. Bases of nimbostratus clouds are wet and ragged. Prepare for long rains and/or deep snow.

If there's enough blue sky to make a Dutchman's breeches, the weather will soon clear.

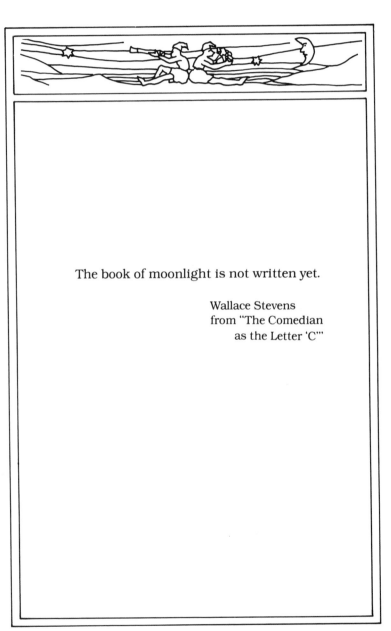

The book of moonlight is not written yet.

Wallace Stevens
from "The Comedian
as the Letter 'C'"

SUN, MOON & STARS CLUES

Clear moon—frost soon.

When the sky seems very full of stars, expect frost or heavy dew.

If the sun goes pale to bed,
'Twill rain tomorrow, it is said.

When you can see the moon's complexion or horns very clearly, strong winds are coming soon.

Excessive twinkling of the stars often means wind is approaching.

If the moon rise pale, expect rain.

The moon, her face if red be,
Of water speaks she. (Zuni)

Referring to the moon's halo:
Circle near, water far;
Circle far, water near.

When the stars begin to hide,
Soon the rain it will betide.

A red sun has water in his eye.

I would like to step out of my heart
and go walking beneath the enormous sky.

Rainer Maria Rilke
from "Lament"

RAINBOW CLUES

A rainbow in the morning
Is the shepherd's warning;
A rainbow at night
Is the shepherd's delight.

If the rainbow's green be large and bright, the rain will continue.

If red is the strongest color, look for wind and rain.

If blue predominates, the air is clearing.

When a perfect rainbow shows only two principal colors, red and yellow, expect fair weather.

If the rainbow is broken in two or three places, or only half of it is visible, expect rainy weather for two or three days.

Seven rainbows, eight days rain.

When the wind is in the north,
The skillful fisher goes not forth;
When the wind is in the east,
'Tis good for neither man nor beast;
When the wind is in the south,
It blows the flies in the fish's mouth;
But when the wind is in the west,
There it is the very best.

Izaak Walton
from *The Compleat Angler*

WIND CLUES

If kites fly high, fair weather is forthcoming.

Do business with men when the wind is in the northwest.

The west wind is a gentleman and goes to bed at night.

If the wind follows the sun's course, expect fair weather.

Unsteadiness of wind shows changing weather. The whispering grove tells of a storm to come.

When the wind's in the south,
The rain's in its mouth.

When the rain is from the east,
It is for four and twenty hours at least.

The south wind warms the aged.

Every wind has its weather.

"A thaw? Those Paris meteorologists can't teach me anything about that! Look at the cat's paws!" Feeling chilly, the cat had indeed folded her paws out of sight beneath her, and shut her eyes tight. "When there's only going to be a short spell of cold," went on Sido, "the cat rolls herself into a turban with her nose against the root of her tail. But when it's going to be really bitter, she tucks in the pads of her front paws and rolls them up like a muff."

Colette
from *Sido*

ANIMAL CLUES

Animals, birds and insects are alert to changes in air pressure and react by flying lower, hurrying home, eating more, dancing or even sneezing.

If the cat is sneezing or dancing or washing her fur against the grain, look out for rain.

A storm is coming when the dog eats grass in the morning and busies himself making holes in the ground.

When a cow tries to scratch her ear,
It means a shower is very near.

When a storm is brewing, rabbits sit quite still by the roadside, ears twitching.

When cows lie down during light rain, it will soon pass.

It will rain if horses gather in the corner of fields.

Before rain, sheep are unusually frisky, leap a lot and bunt or "box" each other.

For, lo, the winter is past, the rain is over and gone;
The flowers appear on the earth; the time of the
singing of birds is come, and the voice of the
turtle is heard in our land.

from The Song of Solomon

BIRD CLUES

Seabirds fly out early and far to seaward at the break of a fair day.

Seagull, seagull, sit on the sand;
It's never good weather while you're on the land.

If owls hoot at night, fair weather is at hand.

If robins are singing high in the tree, the weather will be fair.

Fowls run to shelter and stay there if they think the weather will clear; but if they see it is to be wet all day, they come out and face it.

If the cock goes crowing to bed,
He'll certainly rise with a watery head.

Pigecns wash themselves before rain.

Swallows and larks flying low to the ground indicate rain.

Starlings and crows clustering usually mean rain.

The goose and the gander begin to meander,
The matter is plain—they are dancing for rain.

"You needn't feel too badly, Wilbur," she said. "Not many creatures can spin webs. Even men aren't as good at it as spiders, although they *think* they're pretty good, and they'll *try* anything. Did you ever hear of the Queensborough Bridge?"

Wilbur shook his head. "Is it a web?"

"Sort of," replied Charlotte. "But do you know how long it took men to build it? Eight whole years. My goodness, I would have starved to death waiting that long. I can make a web in a single evening."

"What do people catch in the Queensborough Bridge—bugs?" asked Wilbur.

"No," said Charlotte. "They don't catch anything. They just keep trotting back and forth across the bridge thinking there is something better on the other side. If they'd hang head-down at the top of the thing and wait quietly, maybe something good would come along...."

E. B. White
from *Charlotte's Web*

CLUES FROM INSECTS

If spiders are at work making new webs, the weather will be fair.

A fly on your nose / You slap and it goes;
If it comes back again / It will bring rain.

If many earthworms appear, it presages rain.

When black snails on the road you see,
Then on the morrow rain will be.

When ants travel in lines, expect a storm. If they scatter, it will be fair.

An open ant hill means clear weather; a closed one, rain.

If bees stay at home / Rain will soon come;
If they fly away / Fine will be the day.

If spiders work during rain, it will be a short storm. If they mend their webs between 6 o'clock and 7 o'clock in the evening, it will be a peaceful night. If they leave their webs, expect a storm.

When eager bites the thirsty flea,
Clouds and rain you'll surely see.

Come into the garden, Maud,
  For the black bat, night, has flown,
Come into the garden, Maud,
  I am here at the gate alone;
And the woodbine spices are wafted abroad,
And the musk of the rose is blown.

               Alfred Lord Tennyson
               from "Maud"

PLANT CLUES

Plants, too, are harbingers of weather. Dandelions, tulips, chickweed and clover all fold into themselves before a storm. A strand of seaweed hung up in the kitchen will absorb the moisture in the air before rain.

The silver maple shows the lining of its leaves before a storm.

When grass is dry in morning light,
Look for rain before the night;
When the dew is on the grass,
Rain will never come to pass.

If toadstools spring up in the night in dry weather, rain is on the way.

Pine cones hung up in the house will close up against wet and cold weather and will open when the weather is hot and dry.

The odor of flowers is stronger just before a storm when the air pressure is low.

With dew before midnight,
The next day will sure be bright.

Memorize your house. Then, always pretend you are just visiting. Never expect the spoons to be in a certain drawer, the table, the doors, the bathtub, the teapot, to always be there. Let them surprise you. Search uncertainly for a can opener, wonder if it really is possible that there isn't one, and no milk, no sugar. Never again drop into a chair, bed or couch as if it existed only to cushion your fall day after day. Most of all: don't think the windows will always be in the same place. Touch wood, iron, glass, linoleum, china, silver, cloth, enamel, and whisper a wordless thanks. Never breathe a sigh of relief. This is just the beginning. This is just the house.

Claire Siegel
"Memorize Your House"

HOUSEHOLD CLUES

If coffee bubbles in the center of your cup, expect fair weather. If bubbles adhere to the sides of the cup, forming a ring, expect rain. If they separate without assuming any fixed position, expect changeable weather.

When chairs and tables creak and crack, rain is on the way.

When everything on the table is eaten, look for continued fair weather.

When the odor of pipes and cigarettes is retained longer than usual and seems denser and more powerful, it often forbodes rain and wind.

When matting on the floor shrinks, dry weather is at hand. When the matting expands, expect wet weather.

When boiling water evaporates quickly, rain is near.

When cream and milk turn sour in the night, there are thunderstorms about.

Some rain, some rest;
Fine weather isn't always best.

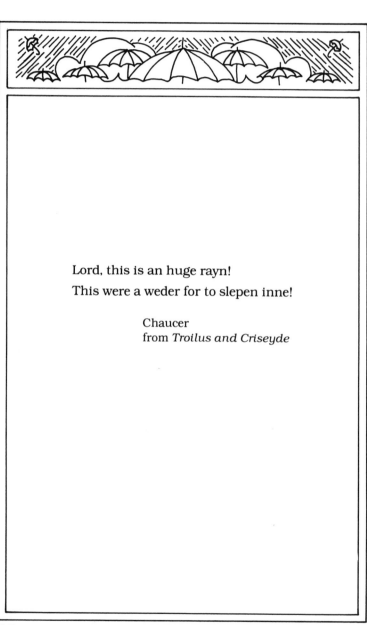

Lord, this is an huge rayn!
This were a weder for to slepen inne!

Chaucer
from *Troilus and Criseyde*

PERSONAL CLUES

Skies are clear, the air is cool and dry, and there may be a pure, sweet waft of breeze. The day tastes like spring water, and for the moment all's well with the world.

Sound traveling far and wide,
A day of rain will soon betide. (Ben Franklin)

You can smell more—good and bad—before a storm.

Your feet and legs may swell when the barometric pressure falls.

A coming storm your shooting corns presage,
And aches will throb, your hollow tooth will rage.

In persons of weak and irritable constitution, the digestive powers are much influenced by the weather. Before storms, such persons are uneasy.

You can expect a weight increase when a cold front passes overhead.

But I know ladies by the score
Whose hair, like seaweed, scents the storm;
Long, long before it starts to pour
Their locks assume a baneful form. (A.P. Herbert)

It was only half-past six, and the minutes of the afternoon were like bright mirrors. From outside there was no longer the sound of whistling and in the kitchen nothing moved. Frankie sat facing the door that opened onto the back porch. There was a square cat-hole cut in a corner of the back door, and near-by a saucer of lavender sour milk. In the beginning of dog days Frankie's cat had gone away. And the season of dog days is like this: it is the time at the end of the summer when as a rule nothing can happen—but if a change does come about, that change remains until dog days are over. Things that are done are not undone and a mistake once made is not corrected.

Carson McCullers
from *The Member of the Wedding*

HEAT WAVE?

When clouds cover the night sky and daytime clouds are a milky haze, the day will be a hot one.

If the setting sun is a fireball and the temperature stays high after nightfall, turn on the fans.

How thy garments are warm, when He quieteth the Earth by the South wind. (Job)

The droning of the "hot bugs" fills the afternoon with sound.

The mailman or mailwoman is wearing shorts.

Your house smells like summer camp, and there is green mold in the closet and mushrooms growing behind the bathroom sink.

High humidity causes doors and drawers to swell, bugs to bite, and joints to ache.

Mirages and tar bubbles appear on the road.

Cats stretch out full-length in shaded spots or on cool kitchen floors.

Snow had fallen in the meantime: one does forget
that annual marvel, the width of it all, the air
given presence, the diagonal strokes of the
streaming flakes laid across everything like an
etcher's hatching, the tilted big beret the
birdbath wears next morning, the deepening in color
of the dried brown oak leaves that have hung on and
the hemlocks with their drooping deep green boughs
and the clear blue of the sky like a bowl that has
been decisively emptied, the excitement that
vibrates off the walls within the house, the
suddenly supercharged life of the wallpaper.

John Updike
from *The Witches of Eastwick*

SNOW?

Three cloudy days generally precede a snowfall.

Look for "snow blossoms"—clouds that resemble white puffs of flowers with a dark gray-blue tint.

The north wind doth blow,
And we shall have snow.

Look for a bearded frost before a snowfall.

The chimney smoke is barely rising and seems to droop, listlessly, into the air.

The needles on the pine trees are turning to the west.

The partridges are perched high in the trees.

The cows are bellowing in the evening.

There is a general animation of man and beast, which continues until after the snowfall ends.

Corn is as comfortable under the snow as an old man under his fur coat.

It will be a long rack of cold, so stock up on coal,
wood, flannels, good books, stewed tomatoes,
matches and a rich, ruby port.

Hannah Rae

HARD WINTER?

A late frost brings a cold winter.

If on the trees the leaves still hold,
The coming winter will be cold.

Carrots and turnips grow deep; the husks on the corn are tight and very thick.

There are lots of layers on the onions—thick ones, too.

There are tough skins on potatoes and apples.

There is a narrow brown band round the woolly caterpillar's middle.

A warm Christmas—
A cold Easter.

If Christmas day on Thursday be,
A windy winter you shall see;
Windy weather in each week,
And hard tempests strong and thick.

March in January—
January in March.

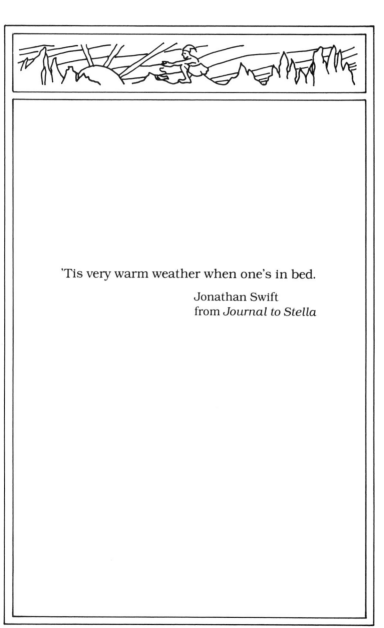

'Tis very warm weather when one's in bed.

Jonathan Swift
from *Journal to Stella*

MORE HARD WINTER CLUES

The nut harvest is plentiful.

Crickets are singing in the house.

Squirrels' tails seem especially fluffy.

Oysters are bedding deep.

There's plenty of honey in the hive.

The skunks are coming in early and making their homes in the barn.

If the groundhog ventures from his burrow on February 2nd and squints at the sun, there'll be six more weeks of winter.

If snow begins in mid-day,
Expect a foot of it to lay.

[In winter] people should retire early at night and rise late in the morning and they should wait for the rising of the sun. (Nei Ching)

Everybody talks about the weather, but nobody
does anything about it.

> Charles Dudley Warner, from an
> editorial in the *Hartford Courant*,
> August 24, 1897

QUICK CALCULATIONS

You don't need a whole host of sophisticated instruments to take a quick fix of the temperature, to determine the "miserability index" of the day's humidity or to figure the distance between you and the oncoming thunderstorm. A few hasty calculations will tell you what you need to know:

THE CHIRPING THERMOMETER:

Male crickets chirp faster when the temperature is increasing. To find the temperature, count the number of chirps in 15 seconds and add 37.

THE MISERABILITY INDEX:

Add the temperature to the relative humidity for a miserability index. If the sum is more than 150, it will be one of those days.

75 degrees + 80% humidity = 155   (WHEW!)
75 degrees + 50% humidity = 125   (AH!)

THUNDERSTORM DISTANCE:

To determine the distance between you and a thunderstorm, count the seconds between the time you see a flash of lightning and hear the thunder. Every 5 seconds equals 1 mile. If, for example, 10 seconds go by, you are 2 miles from the storm's center.

WEATHER CAUSES

We would have no weather without the sun. It is the sun's heat interacting with the earth's sources of moisture that cause the formation of great, bubbling particles of air called "air masses."

Air masses are born over land or sea and take on the properties of their source region. These air masses, once they leave home, are propelled by the winds toward air masses with opposite characteristics. Since warm air moves toward cold and cold towards warm, wherever two opposite air masses collide we have weather.

Collisions take place along "fronts," the boundary separating one air mass from another. When a front is over you, you'll find the air pressure dropping, the winds changing to counterclockwise, the sky scudding over with clouds and often precipitation.

You'll know a cold front is upon you when the air pressure drops, altocumulus clouds appear to the west and northwest and gradually stretch out to stratocumulus, nimbostratus or cumulonimbus, and rain or snow falls. You'll know the cold front has passed when the pressure increases, the wind shifts clockwise, for example, from south to west, the temperature drops, the air freshens and the sky clears.

Warm fronts, which move only half as fast as cold, usually advance gradually with an ever more pervasive cloudiness. Clouds start out with mares'

tails (cirrus), then thicken into a mackerel sky (cirrocumulus) and finally blanket the sky with a gloomy altocumulus or altostratus. Pressure falls, wind may pick up and precipitation occurs. Temperatures stay warm, of course, and humidity generally lingers too.

Changes in air pressure cause all sorts of strange things to happen. When it drops, our own internal body temperature may feel greater and our bones and joints may ache. When air pressure decreases, less oxygen is forced into our systems, and we may feel listless and tired.

High pressure pushes odors back into the earth, and when the pressure decreases those odors are released. Since low pressure usually indicates rain, there are more odors in the air before a storm.

When the air pressure is low, some birds and insects fly lower to the ground where the pressure is higher and flying is a bit easier. Insects swarm and are generally more active in humid weather or when air pressure is low. More bites!

Dr. Edward Jenner, the English 18th-century scientist and physician, appears to have paid almost as much attention to the weather as to his scientific labors (Dr. Jenner created the first successful vaccination for smallpox). He hasn't left a stone unturned in this comprehensive list of rain clues. A list in rhyme no less!

Signs of Rain

The hollow winds begin to blow:
The clouds look black, the glass is low,
The soot falls down, the spaniels sleep,
And spiders from their cobwebs peep.

Last night the sun went pale to bed,
The moon in halos hid her head:
The walls are damp, the ditches smell,
Closed is the pink-eyed pimpernel.

Hark how the chairs and tables crack!
Old Betty's nerves are on the rack;
Loud quacks the duck, the peacocks cry,
The distant hills are seeming nigh.

How restless are the snorting swine,
The busy flies disturb the kine,
Low o'er the grass the swallow wings,
The cricket, too, how sharp he sings!

Puss on the hearth, with velvet paws,
Sits wiping o'er her whiskered jaws;
Through the clear stream the fishes rise,
And nimbly catch incautious flies.

The glow-worms, numerous and light,
Illumined the dewy dell last night;
At dusk the squalid toad was seen
Hopping and crawling o'er the green.

The whirling dust the wind obeys,
And in the rapid eddy plays;
The frog has changed his yellow vest,
And in a russet coat is dressed.

Though June, the air is cold and still,
The mellow blackbird's voice is shrill;
My dog, so altered in his taste,
Quits mutton bones on grass to feast;

And see yon rooks, how odd their flight!
They imitate the gliding kite,
And seem precipitate to fall,
As if they felt the piercing ball.

'Twill surely rain; I see with sorrow,
Our jaunt must be put off tomorrow.

Dr. Edward Jenner

At last count, we found 29 clues. How about you?

47

The winter evening settles down
With smell of steaks in passageways.
Six o'clock.

T.S. Eliot
from "Preludes"

MONTH OF JANUARY

Observations:

MONTH OF FEBRUARY

Observations:

MONTH OF MARCH

Observations:

The sun was warm but the wind was chill.
You know how it is with an April day
When the sun is out and the wind is still,
You're one month on in the middle of May.
But if you so much as dare to speak,
A cloud comes over the sunlit arch,
A wind comes off a frozen peak,
And you're two months back in the middle of March.

Robert Frost
from "Two Tramps in Mudtime"

MONTH OF APRIL

Observations:

MONTH OF MAY

Observations:

MONTH OF JUNE

Observations:

55

Summer—summer—summer! The soundless

footsteps on the grass!

> John Galsworthy
> from "Indian Summer of a Forsyte"

MONTH OF JULY

Observations:

MONTH OF AUGUST

Observations:

MONTH OF SEPTEMBER

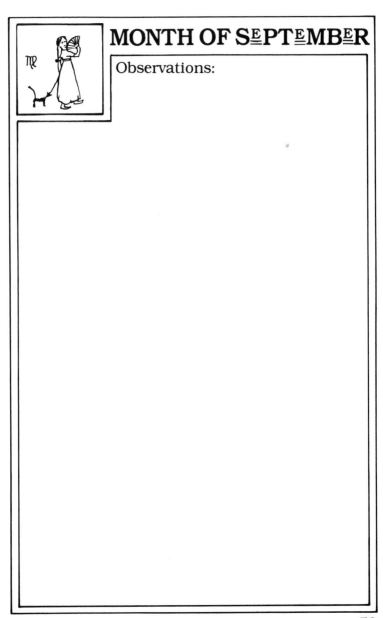

Observations:

I saw old Autumn in the misty morn
Stand shadowless like silence, listening
To silence.

Thomas Hood
from "Autumn"

MONTH OF OCTOBER

Observations:

MONTH OF NOVEMBER

Observations:

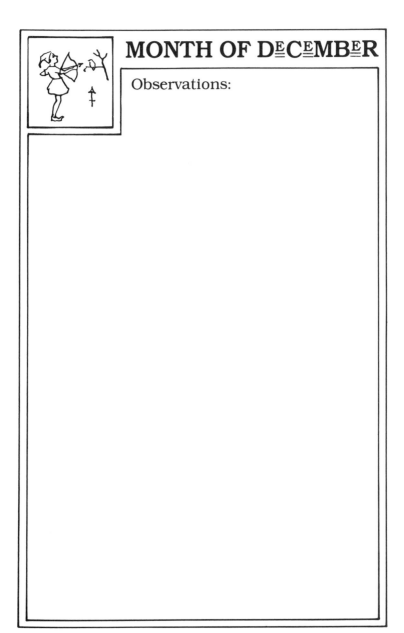

MONTH OF DECEMBER

Observations:

Between two hills
The old town stands,
The houses loom
And the roofs and trees
And the dusk and the dark,
The damp and the dew
  are there.

The prayers are said
And the people rest
For sleep is there
And the touch of dreams
  is over all.

        Carl Sandburg
        "Between Two Hills"